Veterans Reducing Isolation During COVID 19
Vietnam Veterans Diablo Valley 03MAY2022

E5
THERAPY
Where the work gets done!

Veterans Reducing Isolation During COVID 19
Vietnam Veterans Diablo Valley 03MAY2022

Jeffrey Jewell with Matthew Decker

Edited by Don Downey

WWII Veterans History Fund
2022

First Printing: 2022

ISBN 978-1-63010-025-4

Library of Congress Control Number: 2022904830

Foreward

This book consists of the slides and transcription of Jeff Jewell's and Mark Decker's presentation to The Viet Nam Veterans of the Diablo Valley on March 3, 2022.

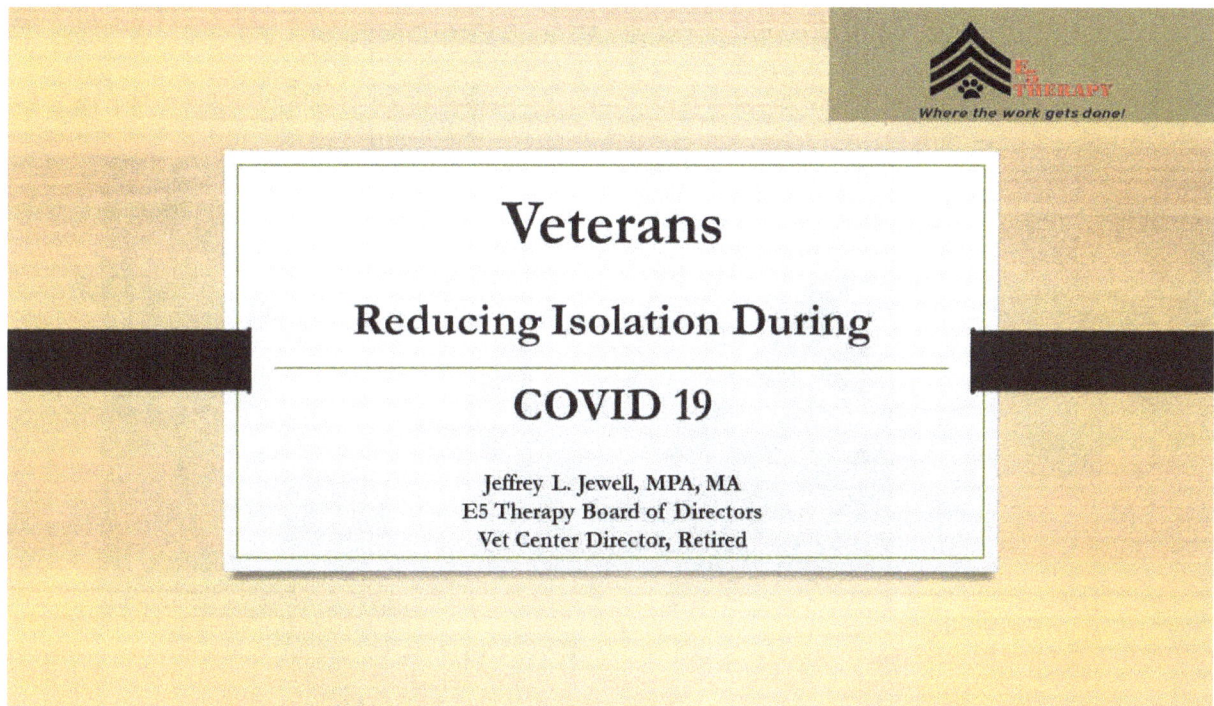

Slide 1: I do want to say, first of all, it's an honor being here tonight. And I know that everyone in this room, like myself as a veteran, that you served not only in your country, but I want to thank everyone of you because I know every one of you continue to serve, because veterans do, especially combat veterans. They come back they have families; they have a job. They pay taxes. They're public servants. They're teachers. They're educators. They're businessmen and they give back to the community. So I want to thank all of you for not only serving your country, but I know that each and every one of you continue to serve. And that's what I want to thank you for. So give yourself a round of applause.

I'm going to do a little presentation tonight. I was recently after I retired, I was contacted by Yolo hospice. They were having a lot of difficulty with veterans in their program. And not only were these veterans in hospice and they're dying, but the social workers and the staff and the veterans... We're all struggling with how to deal with isolation with veterans, because veterans have some really We're going to discuss tonight. They have some real unique issues. So we're going to talk a little bit tonight about how that some of the veterans can cope with the isolation that's brought on by the post-pandemic world.

And I got a little bit of help tonight. Because part of the section of this is we're going to talk a little bit about pets and how pets can impact and help reduce blood pressure, help reduce isolation, depression. And I want to introduce to

8

Matt Decker and Larson. Once you come up here and when I get that little section, I'm going to have Matt talk a little bit about service dogs and how pets can help and what we call it medication without side effects.

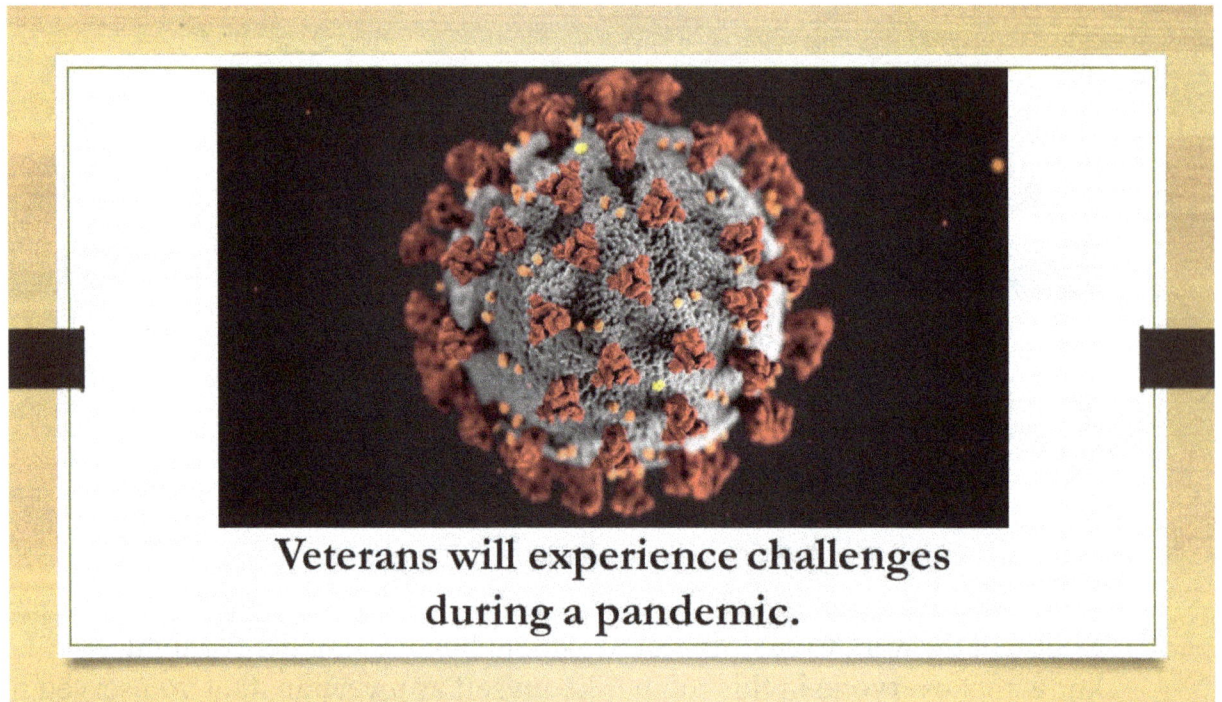

Veterans will experience challenges during a pandemic.

Slide 2: I had to search really hard for this next slide. This is the actual first picture of the coronavirus. This is the one that came out of China, the original one right there. So veterans who will experience challenges during the pandemic.

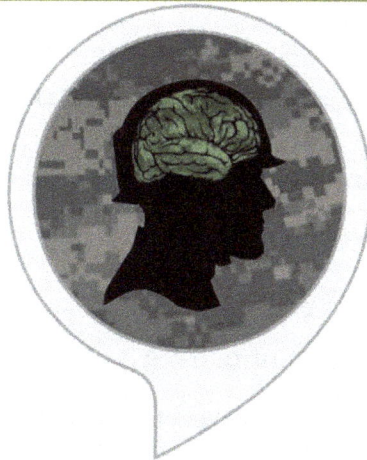

Slide 3: Some Veterans will struggle with mental health issues.

Slide 4: All of them certainly have earned our help.

Common stereotypes about veterans:

- Veterans are angry
- Veterans are men
- All veterans are in crisis
- All veterans can obtain VA services
- All veterans have served in combat
- You have to be in combat to "get" PTSD

Slide 5: Extremely important that we talk a little bit about before we do anything else. As I spent my entire career with the VA and as a veteran service officer, the last 35 years, talking about veterans and trying to dispel some of these myths and stereotypes that the media and TV and everybody portray, how they portray veterans, especially Vietnam veterans. So the stereotypes I hear all the time - *Veterans are angry*. Nothing could be further from the truth. *All veterans are men.* Bullshit, women served. One of my last acts as a director of the Concord Vet Center is I had 300 women veterans, the American Legion Hall in Vallejo and celebrated their service from all over the Bay Area. Women serve too and some, are a lot of them are serving in combat roles now. *All veterans are in a crisis.* This is what the public assumes about all of you. A good majority of the public and *All veterans can obtain VA services.* Unfortunately, that's not true. I'm one of those few dinosaurs that worked in the Veterans Health Care Administration, the vet centers, the National Cemetery System, sat on the board. And I also did benefits and worked with the Veterans Benefit Administration. I'm one of the few dinosaurs that understands all the entities within the VA. And when people just say, oh, you can go to the V.A. and that's not really true. It happens. And everybody is unique. Everybody's different. And there are a million different factors involved in veterans VA services. It is by far the most complicated bureaucracy, way above Social Security. Social Security is second, V.A. bureaucracy is top of the list. *All veterans served in combat.* That's not true. You have to be in combat to get PTSD. Bullshit. Excuse my language. I worked with a lot of veterans that had, saw some pretty horrible accidents In peacetime that had post-traumatic stress disorder and suffered horribly. And military sexual trauma. One of the worst things that had to do is the counselors worked with men and women that were severely traumatized in the military through sexual trauma.

Positives of Military Service

- Pride
- Values and honor
- Significant responsibility, especially during war time
- Competency
- Sense of accomplishment
- Sense of meaning and belonging
- Development of close relationships/family
- Cross-cultural experiences

Slide 6: So if we're going to talk about stereotypes, we need to talk about some of the positives, positives of military service, pride, values, honor, significant responsibility during wartime. So guess what? Why wouldn't you want to hire a veteran? I had a TV reporter in Sacramento one time to try to get me to say the Vietnam veterans with PTSD can't be employed. He tried like hell for a half an hour, holding that microphone in front of me. Every time he tried to get me to say something negative about a Vietnam Veteran I'd say, well, why wouldn't you want to hire a veteran? Gee whiz, they show up on time. They take responsibility for their actions. They're good workers. They're committed to the community, you know? And so every time you try and get me to say something negative, I tell you. And God bless him. And I watch all this, all the interview. He didn't say one negative thing about Vietnam better, but he sure as hell tried to get me to say something on camera. Veterans are competent. They have a sense of accomplishment, a sense of meaning, sense of belonging. And that's why they that's why they work so well in the community. And that's why they volunteer at the churches, and that's why they volunteer at their school. That's why they're going to guys like Bill Green here. Oh, my God. How many students? Bill Green: 90 some thousand. Thousand students standing up and talking to them, talking to them about Vietnam and the military and military service. Sense of meaning and belonging. They have a sense of meaning, development of close relationships and family, cross cultural stuff. Oh, my God. I'm not going to lie to you. I grew up in all lily-White, Green Bay,

Wisconsin, the only African-Americans when I was in Green Bay, Wisconsin, in the 1950's and 60's and 70's, early 70's played for the Green Bay Packers, not kidding. [light laughter] the local police chief. No, no kidding. He used to, any African-American, he put him on a bus to Milwaukee, 120 miles north. I mean, that's what I grew up in and I think one of the reasons I actually wanted to go in the military not go right into college is because I knew that there was something beyond lily-White Green Bay, Wisconsin. And I am glad I did and I'm willing to admit that I grew up in that environment. But now after my cross-cultural experiences, and my life experiences in the military with people from all walks of life has made me a better person. And it given me a unique perspective about the world in which every one of you have in this room.

How do we keep from isolating
during a pandemic?

Slide 7: How do we keep from isolating during the pandemic? Okay, let's talk about that.

Buddy Checks

Connecting with family, other veterans or friends.

Slide 8: Buddy Checks. Connecting with other veterans and friends.

National Institute for Occupational Safety and Health

NIOSH

What is the Buddy System?

The buddy system is an effective method by which a deployed staff member shares in the responsibility for his or her partner's safety and well-being. This type of active support is important in any deployment. Buddies are responsible for looking after each other in two main areas:

• Personal safety
• Resilience

Slide 9: So how do we do that? The best model that I could find when I was doing my research comes from the National Institute of Occupational Safety and Health. I think that they had the best description of a buddy system. The buddy system is an effective method by which we deploy staff members share in the responsibility for his or her partner's safety and well-being. I wanted to really define it, so how do we how do we do those buddy checks and what do we do? What are some of the things we say? What are some of the things we shouldn't say? That's what we're going to talk about.

Putting the Buddy System into Action
- Deploy in 2-person teams (minimum).
- Look out for hazardous conditions, safety demands, and stressors.
- Manage stress to prevent burnout.

Slide 10: Putting a buddy system into action. It was best to deploy the team leaders and folks with the American Legion, it seems to be the one veterans service organization that led the way nationally right off the bat during the pandemic. They were immediately going to their 2 million members nationwide, and they were saying, look, every post, every district, every department in California or wherever you're at, they were asked if they were saying, look, you need to be reaching out to your members. You need to be calling them. You need to visit them, make sure they're So these are some of the American Legion folks that were right off the bat, they're checking in with members.

Buddy System Actions

Do

- Be a listener to your buddy.
- Actively communicate with your buddy to understand his/her perspective.
- Reach out to a buddy who may be struggling.
- Offer help with practical needs or finding services.
- Get help if you have reason to believe your buddy may be a threat to themselves or to others.

Do Not

- Offer clinical diagnosis or treatment.
- Take on the role of a therapist.
- Pass judgment on people or decisions.
- Pry or demand that a buddy discuss problems.

Slide 11: Buddy System Actions. Some of things to do and not to do when we're checking on somebody that you think might be depressed or by the way, if you think somebody let me just say this. Don't ever be afraid to ask somebody if they're thinking about suicide. If you even have an inkling in the back of your head they're thinking about suicide. Do not be afraid to ask the question. Let me tell you why. Because 99% of the time, if you think somebody is depressed, suicidal, you know, the hair in the back of your neck stands up. And I know it does for every one of you guys in this room, you need to ask that question because I cannot tell you how many times in my career where I've sat with a veteran and I said, hey, Larry, you're thinking about suicide? Absolutely. Oh, my God. Nobody's ever asked me, so if you think somebody is, you know, acting a little weird, maybe he's given it away to all his friends. Maybe he's having too much to drink, ask the question, because if you truly mean it, they're going to tell you and then you can get them some professional help. So be a good listener to your buddy. Actively communicate with your buddy to understand their perspective. So what does it mean to be a good listener? Anybody tell me. Someone in audience: You're awesome. Yeah. It's hard for me. I'll be the first to admit it as a therapist, you know? I mean, there are... several people even there are two or three of you in this room. I'm going to tell you right now that 15-20 years ago, you walked into my office and I never met you before. And within 2 minutes, I knew you had full-blown post-traumatic stress disorder. You really needed help or where. And I had to bite my tongue to keep from saying you're pretty screwed up. You know me here because I knew that if I said something, there and there, you know, guess what? Joe's going to run. Never come back. So being an active listener is not only listening to what they're saying, but being able to give them the information in a way that they can understand it. Because ultimately I'm not the one that needs to change their behavior or feel better if the person sitting across from me. And so my job is to facilitate and teach that person how to help themselves. And I'm never going to work harder than that person sitting across me. And if they're if they don't think you're invested in them getting better they're not going to go with you. So and I've had that's six months, a year, two, three, four years. They would finally say something and I would go and I'd knock my head off my head just like that. And I go, Oh, my God, what did you just say? And I would repeat it back to them because in that moment it's been five years. They finally said the one thing that's going to help them heal. So that's what that's what it's really all about. Get help. If you have reason to believe your buddy may be a threat, you're not a therapist, but if the hair on the back of your neck stands up let's get them some professional help. What not to do, offer clinical diagnosis or treatment, hey... what not to do as a therapist and I made that mistake - was don't diagnosis family and friends because that'll get you in a lot of trouble. And never and never give your wife an IQ test. When I was getting my master's, they sent me home with the IQ kit, the kit and the instructor said, don't do family members. I did my neighbors, but I had to do one more person. And I had class the next day. And so I did the whole test with my wife. I'm going, holy shit, she's smarter than me, she's going to score off the scale. When I scored this thing. And so I got to the end and I went, okay, thank you very much. You go now you're to score it, right? I go, I wasn't going to score you. I knew I was in trouble. So you never diagnose or treat family or friends. That's a good thing to remember, especially as therapist. Don't pass judgment on other people, just by the way they look, by the way they talk. Really try to understand where they're coming from. And here's something. When these veterans come to me and I knew they had PTSD, why would I want to talk about PTSD in the first visit if I know the veterans got post-traumatic stress, I know he's going to have trouble if he's not sure where his next meal is going to come from. He's going through a divorce. He doesn't have enough money, not sure where he's going to live. What are we going to do? We can address those underlying issues before we even get to what's really going on with the veterans. Don't try to pry things out. If somebody is uncomfortable, tells you to back off, you know, go ahead and back off.

Pets – A Medication Without Side Effects

Slide 12: I'm going to turn the microphone over to Matt Decker. Matt is used to be a therapist at the Concord Vet Center. He also worked, what, three or four years and the clinical director for the Purple Heart of Santa Rosa. I'm on the board of E5 Therapy up in Suisun. There is no vets center up at Sloan County. And we have the largest population in the country of returning veterans. And so Matt started a nonprofit. And we have totally outstripped in two years our seven-year projection. And we're seeing hundreds of veterans and we've got how many therapists now? Matt: Five, five therapists. And if you can't afford it, if you don't have any insurance or whatever, we're paying for it. We're nonprofit. And so I'll have Matt talk a little about "Pets Medication Without Side Effects". Matt: One correction. I was not the executive director of Paws for Purple Heart. I was a clinical director but I am the executive director of E5 Therapy. The goal there is to fill the service gaps that the VA has. We don't limit ourselves to who, what, where, when, why and how we do whatever it takes. The model of E5 therapy, it's where the work gets done. How that happens is up to the veteran that walks through the door or the spouse or their child or anybody that lives in the house with them. As people of service, we don't just sit within ourselves. We have families we have friends. And if they come to us and they say, Hey, my wife's struggling, great, let's help her. Why? Because that's the best way I can serve that veteran. I'll serve the wife, I'll serve the husband, we'll do the couple's work, we'll see the kids, whatever the veteran needs, we're going to fill that gap. Places the VA isn't going to go right now. Now on to the fun part, pets.

Pets – A Medication Without Side Effects

- Physical
  - Research proves that in the presence of a pet dog, the human counterpart experiences a decrease in blood pressure and heartrate.
  - This physical change increases as the human engages with their canine counterpart.
    - Petting, sitting in contact, cuddling, or engaging in play.

Slide 13: Pets are literally a medication the same as all of those drugs that the VA gives out for depression and anxiety. We can find the same benefit and clinical impact utilizing well behaved and trained dogs and yeah, if we had to... cats. I bet you can guess which side of the fence I land on. [audience chatter] Funny thing about this is when I started my work in canine assisted therapy, I had to say things like research suggests since the time I started in canine assisted therapy, we've actually gotten the authority and permission to say research now proves which is huge for us because suggesting that research is a good thing. Research proves that the presence of a pet and their interaction with the human counterpart decreases their blood pressure. And their heart rate. Now, I know it sounds a little weird. Let me get more specific. When I say blood pressure and heart rate. 30 pets per minute. That's you sitting on the couch going.... Decrease in heart rate, decrease in blood pressure. The physical change increases as you engage with your pet. You'll see a decrease in stress hormones like cortisol. You'll see an increase in serotonin. ... That's what those antidepressant chemicals are all trying to build up in your system. You'll see an increase in oxytocin. You'll see an increase in dopamine.

Pets – A Medication Without Side Effects

- Chemically
  - Research proves that the stress hormone Cortisol is measurably and substantially decreased by interactions with dogs.
  - The release of Oxytocin is almost doubled when talking, stroking, and eye gazing with a pet.
  - An increase in other chemicals also occurs: Serotonin and Dopamine.

Slide 14: Chemical impacts. We've got this idea that we can utilize the pet and gain the same chemical impact as an SSRI or an anti-anxiety because of something that we've learned in the research called synchronization. Synchronization is a pack behavior. As I'm sure most of you know, we are pack animals just like the dog. And because we're a pack animal, when we are engaged with our pack, we will start to calm down and we will start to engage with our pack the same way they're engaging with us. That means as the dog calms down, which is why it's really important not to have Chihuahuas but as the dog calms down, so will you. It's natural. It's biological. We've spent depending on the research. You look at 40,000 to 400,000 years domesticating the dog, their brain and our brain proportioned differently because they smell way better than we do, are almost identically mapped the way a dog recognizes the face of a loved one is the same way your child does. That's how ingrained this is. It's not something that can be resisted, sure, you want to sit there and get all amped up while you're cuddling a 75-pound lap dog? Sure you can get yourself amped up, but if you give yourself over just to being relaxed, the dog's not going to judge you. A dog's not going to judge you. They're not going to cheat you. They're not going to steal from you. And without all of that judgment and without all of that pressure, in your head, all of a sudden, you know they will not steal your dinner. I heard that. They might be a little opportunistic. In the Marines, we don't lie, cheat or steal, but we may acquire some things every once in a while, specifically, from our Army friends. But if you can imagine, for those of you that take the VA cocktail why not try this at home? Can take a couple of deep breaths with your favorite four-legged friend, get 30 pets permitted in, if you have one or more Chihuahuas, this could be very difficult. Do not chase them. Just bring their food to your lap. Works every time. But this can drastically improve your issues with isolation. It can drastically improve your issues with motivation. In all of those positive chemicals that get suppressed over time. Because we've dealt with a lot of crap, haven't we? We want to build those chemicals in... your SSRIs, your antidepressants, your anti-anxieties all have side effects. Canine assisted therapy and the practice therein is quite literally a medication with no side effects whatsoever. Thanks for your time. Audience: Oh, yes. Jeffery: Matt's going to do a short demonstration here. Matt: Yes, you're a good boy. Yeah. Come here. Come here. Can you give a snuggle? Yes, this is good boy. Yes, this is good boy. 3 one thousand, 4 one thousand, 5 one thousand. How easy is this? And if you sit here and pet on the dog they're not going to leave because this is their best therapy. It is 100% our best therapy because of stress because this is this is where you don't have to worry about the judgment and the love. It comes from here no matter what. It doesn't matter if you yelled at them 5 minutes ago, from chewing on your shoes, they come back way faster than any wife after you yelled out to them. [heavy laughter] All right, my wife. Jeff: Thank you very much. [applause]

Slide 15: The challenges veterans face when isolated.

Challenges Veterans Face When Isolated

- Feeling separate - As veterans we are different

- Managing expectations – I can make it through this

- Letting down "guard" – Allowing ourselves to feel

- I have to be strong – I am a veteran and we survive

- Reintegration– Connecting to old friends

- Lack of daily structure – What should I do today?

- Suffering from invisible wounds - There are many types of injuries

Slide 16: Feeling separate. Veterans often feel different. If you've been in a combat situation, guess what? You could not have been affected in some way. So in my experience, veterans, whether you are in combat or not, you look at the world differently and it's because of that whole life experience. Maybe it wasn't in combat maybe it was with your job and whatever branch or service period or whatever. It could be anything but veterans tend to look at the world differently, and they often think that others around them don't understand the world. So the challenges they face, letting down their guard, allowing themselves to feel. Veterans are always going, I have to be strong. I'm a survivor. They don't like to ask for help, you know, reintegration. They have problems connecting with old friends, lack of daily structure. What should I do today? Suffering from invisible wounds of many types.

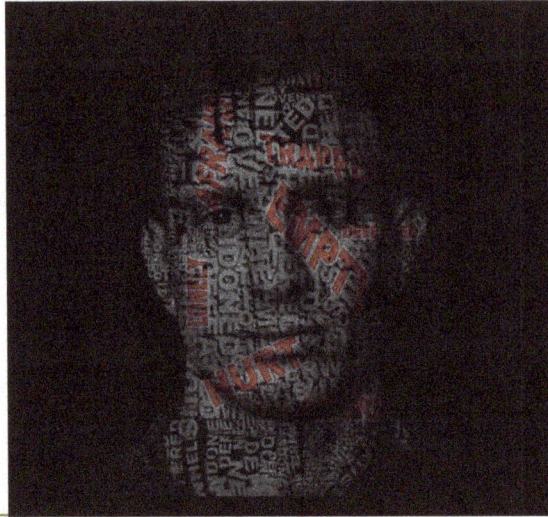

Slide 17: Other factors impacting veterans.

Slide 18: Situations remind veterans of past events. Most veterans trying to avoid situations that are people or memories And this can increase the isolation. So I can imagine someone in hospice right in there isolating because COVID imagine a veteran who has post-traumatic stress disorder and some of the issues that compound the problems that he might have in this post-pandemic world.

Feeling numb: Many veterans find it hard to have positive or loving feelings toward other people and may stay away from relationships.

Slide 19: I love this photo because I see this young lady here has the 'thousand-yard stare". Feeling numb. Many veterans find it hard to have positive or loving feelings towards others. And so this is another issue that we ran into with veterans with the isolating and their anxiety and depression is because because they have a really hard time expressing their own loving feelings. And for any of the wives out here, they can probably stand up and tell you a lot of stories home.

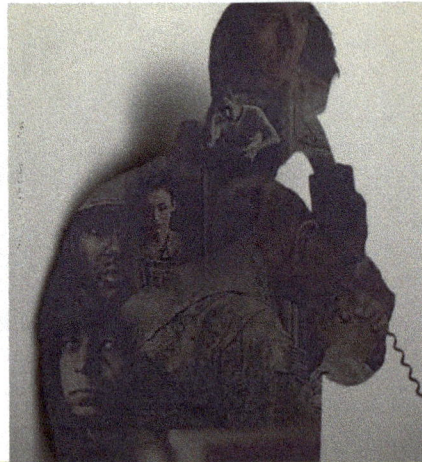

Slide 20: Feeling jittery can cause you to suddenly get angry or irritable, sleep problems. Concentration impaired. Always feeling on guard. Easily startled.

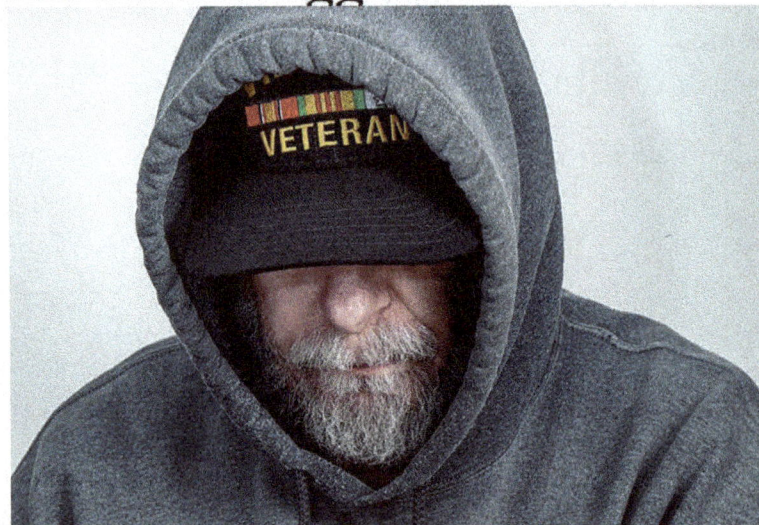

Slide 21: Isolation triggers for veterans.

Smells

Slide 22: So very important to understand that veterans, especially combat veterans, may react to our five senses. Which is smells. Now, my favorite, I had a search to find a photograph of diesel here I hate to tell you how many times in therapy I sat across from Vietnam veterans I'll never own a damn diesel truck! And I go well, why is that? Audience member: Because I had to burn shit. Jeffery: [repeating] Because I had to burn shit. so the smell of diesel triggers those memories from all the diesel trucks from having to burn shit, from all the diesel fuel and everything that was used in the combat zone. So smells can be a trigger.

Sounds

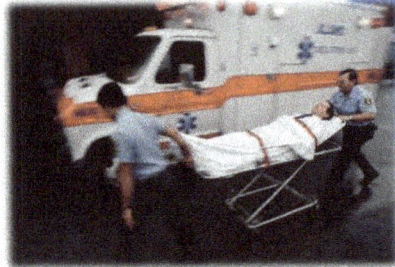

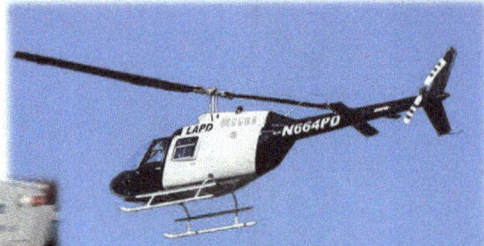

Slide 23: Sounds. Of course. My favorite one, a lot of the 4th of July. Oh, my God. And almost every veteran I've ever seen. What do they do on the 4th of July and New Year's Eve? They go into their man cave and they put it on their headset and turn up the music and watch TV, or they turn out the news rations they don't have to listen to what? The fireworks or the gunshots.

Anniversary Dates & Holidays

Slide 24: And this is one that a lot of folks tend to forget, especially I'm very cognizant of this, especially in all the work I did with bereavement that anniversary dates could be a real trigger. And, you know, Roxanne, back there

is a Gold Star Mom I think I told her one time, I said, Roxanne, you know for the next 18 months, you're going to struggle. Why? Because you going to come up upon Thanksgiving, Christmas, your son's birthday. The day he went into the service, maybe it's his favorite holiday or whatever it might be. So usually if you have a death of a close loved one, there's that 18-month period. But anniversary dates could also be a trigger for a lot of veterans. And we have to understand this. Vietnam veteran in my office maybe 20 years ago, I when I first started at the Oakland Vets Center and he comes in and he goes you know, you guys, December, January, it's January. I think it's in late December or early January every year, man. I'm depressed. I can't figure out what's wrong. I can't sleep here. And after really talking to him, we finally figured out that every year around the Tet Offensive because he was there, audience member: You know it. and then I brought his wife in and his wife goes, holy shit, that's it, you know? So once we opened that box and he understood that this is the reason why he was so depressed and angry and isolating himself and mad at the kids. And man and his wife were like a whole month. Once we understood that, then we were able to work with that and we were able to move beyond that. I still get a card, Christmas card from this guy's wife. Every year telling me how I saved his marriage just because we brought her in. I brought her in and we had a conversation and we and I had the veteran really talk to her, and we figured out what what's really going on here? What's the underlying problem? Anniversary days. Huge trigger for a lot of folks. And I know that many of you in your thinking of a particular anniversary day.

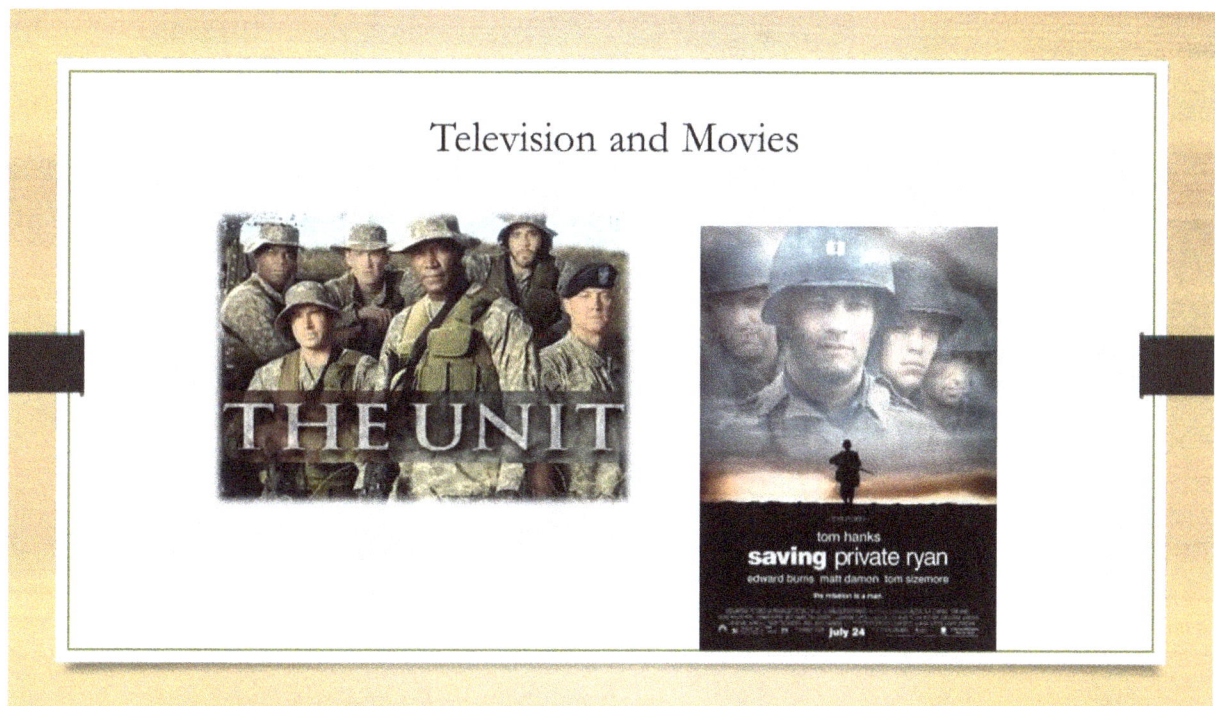

Slide 25: Television and movies. Oh, my God. My wife used to get so pissed at

me. Every time we have a movie come out, I think the last one was Hacksaw Ridge or whatever I say, "Hey, I got to go to the theater. I got to see it when it comes out." and she goes, well, why? And I said, because I'm going to my Vietnam Veterans Group next weekend and I'm going to say to them, you, you, you, you don't go see that movie because it is going to be a huge trigger for you and we're going to set us back in therapy. So I used to go out and try all the movies, especially those with a lot of combat scenes in them and everything. And you really have to be cognizant of that because it could be a huge trigger. It's like Saving Private Ryan.

Slide 26: How can we help?

Asking Questions

Good ways to start a conversation include:

- What did you do in the military?

- Where did you serve? (Don't assume all veterans served in Iraq/Afghanistan.)

- How are you and your family doing?

Slide 27: Start a conversation What did you do in the military? Where did you serve? Don't assume all veterans served Iraq, Afghanistan, or they assume somewhere else. Don't assume that a woman.... Well, I was telling a story earlier, my first day at the Concord Vet Center a young lady walked in and the guy at the counter said "Are you here for your husband in group?" And I went, holy shit, I walked out there and I said, young man, come here and I said, I looked at this lady. You walked in and I said, "Did you serve?" And she said, Absolutely. Thank you very much. And I brought the entire staff in on the first day, brought everybody out of the office. And I said, "first person to say to a woman when they walk through these doors, Are you a wife or spouse of a veteran? You're fired." Hey, when they when women walked through the door, the first thing you need to say is, "Did you serve?" $ Because if you say that to a woman veteran, they're a hooked, they're there. $ If you don't, they're going to leave? And so we were very successful because we appreciated and respected the fact that women served. How are you doing with your family? Good question to ask.

Asking Questions

You should avoid:

- Pressuring a veteran regarding specifics about their service.

- Minimizing the challenges a veteran might face.

- Making assumptions about any veteran's political or foreign policy views.

Slide 28: Things you probably shouldn't do if they don't want to give you specifics, let it go. Don't ever minimize things because minimizing is terrible. We're all really good at minimizing our own stuff and it's really bad if someone else minimizes things in your in your life. So don't minimize things in a veteran's life. Making assumptions about any veterans. Political or foreign politics does not belong in any discussion. And for any of you that may have anything in my group, guess where well, when politics came up, what did they should do? Stop. You got about 10 seconds. We're done. If you let it go longer than 30 seconds you're always going to have one guy extremely to the right and one extreme, the left and a whole bunch of people in the middle. And I feel that politics take over a conversation. And then in a group, guess what you get, you got fisticuffs going on, which happened a couple of times, by the way, I got pretty good at breaking up fights.

Recommendations

- **Maintain awareness** of issues that may impact veterans.

- **Remember that mental health & physical issues** can make it difficult for veterans to understand some processes.

- **Avoid putting the veteran on the spot**, even when veteran appears comfortable doing so. *Not everyone is ready to talk.*

Slide 29: Maintain awareness of issues that may impact veterans. Remember, the mental health, the physical issues may make it difficult for some veterans to understand things. So not only do we have physical issues, we have other mental health issues that we have to be aware of. Avoid putting the veterans on the spot. Even when a veteran appears comfortable doing so, not everyone is ready to talk. But just say, hey, how are you doing? Where's did you serve? Tell me about yourself. One of the things that I learned in bereavement, especially with the parents of the family members, is I had to really bite my tongue. because it really wasn't me. I was uncomfortable. Most people are uncomfortable when they talk to folks who have lost a loved one. But my favorite thing to say was, Tell me about your son. Tell me about your daughter. I had many times where I felt real uncomfortable. And they said, hey, you know, come on, I want to I want to have coffee with you. And I really realize that what they're really saying is, come on over here. And I want to tell you about my son when we finally figured that out believe me, it became pretty easy.

Recommendations

- **Be flexible** whenever possible with a veteran. Sometimes they can't tell you what's *really* bothering them.

- **Encourage involvement** Many veterans lead isolated lives. Sometimes a little (sensitive) encouragement from someone they trust can make all the difference

Slide 30: Be flexible whenever possible. Sometimes you say you can't tell what's really bothering them. Encourage involvement. Many veterans lead isolated lives. We all know what all what you guys do. You encourage others to become involved. They're pretty good to get people involved. Slide 31: Remember, the characteristics of a veteran, resiliency and strength, life experiences equals diversity, motivation and determination, leadership and maturity. And I've got to tell you, from a guy from all lily-white Green Bay, Wisconsin, the Air Force, the military was the best thing ever for me. It really opened my eyes to the whole world around me. I honestly felt if I were to go on straight to college and go all my brain and just I probably would have fallen flat on my face. I didn't go back to school till I was 33. Then I had my last master's degree on my 30th birthday and I probably would fail if the military hadn't given me those life experiences and gave me the courage and the responsibility and the understanding of the real world.

Remember, when you are not
sure what to say to a veteran, just
say…

Slide 32: Remember when you're not sure what to say to a veteran just say what?

Welcome Home.

Slide 33: Welcome. Welcome home. And it's my presentation. Thank you very much, Sir. You have been really fortunate this is your chance now. You know this is your chance. I'm all I don't say this too often because after I retired, I started playing poker with a group of guys where nearly all of them were veterans, because it is somebody like me that's worked with veterans all my life. Every time I turn a corner, somebody had a question about VA benefits,

medical care. And so when somebody asked me to join this little poker group on Thursday night, I said, is anybody a veteran? And they go, No. I said, I'm in because I can see it. I can laugh, I can talk to you. And they're not asking me about my VA compensation or the VA screwed me on this man. So this is your chance. I'm giving you a chance. An opportunity. This is the one of the few dinosaurs here that understand all the vanity. So if you have a question that you think everybody out here could benefit from, I can maybe answer if I can't, I'll find your answer. You're all good. Oh, my God, no. So nobody's ever had an issue with every competition, right? Nobody ever had an issue with a doctor in the area. I don't know. Actually, I will say this the department I'm very proud of the fact that I worked for the VA. And I have to tell you this, and I truly mean, this is 99% of the VA employees and I see day in and day out, they work miracles. Is that one or 2% that you see in the frickin' news all the time that they steal money and deny veterans access or whatever bars I return, they should fire it at one or 2%. But I'm telling you, the majority of the VA employees, I have witnessed miracles day in and day out. And what do they do they move on to the next veteran and they help the next one. And they don't they don't take a pat on the back. They're not out there bragging about it. They're not out there talking to news people or anything. They're just doing their job day in and day out. And they have they have improved. They're not perfect, but they've improved substantially in the last several years compared to the 1940's, 50's, and 60's. So take advantage of it.

VNVDV Member: Thank you very much. I would like to thank you for taking the time out to speak with us today. I thought was very I learned a lot so but it's a really important job and I really appreciate all the all the efforts you work to put this presentation together. I'm really happy about Larsen and I thought that Matt did a wonderful job with his portion of the presentation, like to present to you on behalf of the Vietnam Veterans of Diablo Valley. A little liquid refreshment. Where do you think when you have a little taste I think of us.

Jeffrey: Thank you very much. Yeah. And I really again, I when I put this presentation together, but I called Matt and said, man, give me some help because I want to talk about pets. And so most of that information about pets being the medication without side effects came from Matt and all his work with Purple Heart and as a service officer. Thank you Matt and Larson and *thank you all for taking time to be here tonight.*

"E5 Therapy provides mental health services to anyone willing to put in the work for wellness. Our specialization is canine-assisted therapy for military Veterans. Our leadership has been in practice for more than 10 years, helping people accomplish their goals while fighting for emotional wellness. At E5 Therapy we don't hesitate to work hard and you shouldn't either." Matthew Decker, LCSW #83925, is the founder of E5 Therapy. He was born and raised in Phoenix, AZ and immediately after high school graduation he joined the United States Marine Corps. While deployed to Iraq during Operation Iraqi Freedom, he realized how much his Marines needed emotional and mental health support. Returning home, he completed a Bachelor of Sociology degree and a Master of Social Work degree. He is now a license clinical social worker in Northern California, focused on helping Veterans achieve their mental health goals. E5 Therapy 501(c)(3) use PayPal to donate or if you prefer to send a check, you can mail it to: E5 Therapy 333 Sunset Avenue, Suite 110. Suisun, CA 94585 100% of all donations received are used to directly subsidize the cost of our counseling and training programs for our Veteran clients.

From https://e5therapy.org

Works Cited

Center for Disease Control National Institute of Occupational Safety and Health Buddy System Fact Sheet- https://www.cdc.gov/niosh/index.htm

E5 Therapy Matthew Decker, LCSW - Dogs – A Medication With No Side Effects – Power Point Presentation

References for Veterans

- E5 Therapy -707-225-7899 https://e5therapy.org/
- Advise Nurse N CA – 800-382-8387
- VA Crisis Line – 800-273-8255 Press 1
- VA Vet Centers – 877-WAR-VETS https://www.vetcenter.va.gov/
- Paws For Purple Hearts - https://pawsforpurplehearts.org/

References for Veterans

Assist Dogs International - https://assistancedogsinternational.org/

Bergin University Canine Studies - https://www.berginu.edu/

Image Credits

Slide 2, Food and Drug Administration, Food Safety and the Coronavirus Disease 2019 (COVID-19), 1/27/2022, retrieved March 3, 2022, https://www.fda.gov/food/food-safety-during-emergencies/food-safety-and-coronavirus-disease-2019-covid-19

Slide 3, Amazon Images, Stock Photo, Unknown date, retrieved March 3, 2022, https://images-na.ssl-images-amazon.com/images/I/71-0knwlq2L.png

Slide 4, Mohave Community College, Military Services, Unknown date, retrieved March 3, 2022, https://www.mohave.edu/resources/veteran-and-military-services/

Slide 7, Open Access Government, Veterans are experiencing loneliness and social isolation, December 14, 2018, retrieved March 3, 2022, https://www.openaccessgovernment.org/veterans-experiencing-loneliness-and-isolation/55738/

Slide 8, Imgur, Buddy Check 22, Oct 22, 2015, retrieved March 3, 2022, https://imgur.com/gallery/oFgb5ZM

Slide 10, American Legion, American Legion joins bipartisan effort to pass Buddy Check bill, Oct 8, 2020, retrieved March 3, 2022, https://www.legion.org/legislative/250663/american-legion-joins-bipartisan-effort-pass-buddy-check-bill

Slide 12, Forbes, The Biggest Trends In The Pet Industry, Nov 27, 2018, retrieved March 3, 2022, https://www.forbes.com/sites/richardkestenbaum/2018/11/27/the-biggest-trends-in-the-pet-industry/?sh=61d58ffff099

Slide 15, KHQ News, Veteran suicide prevention bill aims to improve mental healthcare for veterans, Jan 16, 2021, retrieved March 3, 2022, https://www.khq.com/regional/veteran-suicide-prevention-bill-aims-to-improve-mental-healthcare-for-veterans/article_ab6f17f8-4184-5ac9-9726-1698ff22dd46.html

Slide 17, forthoodsentinel.com, Stock Photo, Unknown date, retrieved March 3, 2022, https://bloximages.newyork1.vip.townnews.com/forthoodsentinel.com/content/tncms/assets/v3/editorial/3/c8/3c801812-1e0e-11eb-ad62-d306b52f6993/5fa1b535ccc8e.image.jpg?resize=1200%2C1553

Slide 18, Amazon Images, Stock Photo, Unknown date, retrieved March 3, 2022, https://m.media-amazon.com/images/I/41Q1B3vMfML._AC_.jpg

Slide 19, Elite Learning, Stock Photo, Unknown date, retrieved March 3, 2022, https://www.elitelearning.com/wp-content/uploads/2018/04/Vet900x500.jpg

Slide 20, HBO Films (Amazon Images), Crisis Hotline (video), Unknown date, retrieved March 3, 2022, https://www.amazon.com/Crisis-Hotline-Veterans-Press-1/dp/B00KF8L880

Slide 21, Colombia University (iStock), New Evidence that a Brief Form of Therapy Can Help Veterans Adjust to Civilian Life, November, 2020, retrieved March 3, 2022, https://www.tc.columbia.edu/articles/2020/november/a-brief-form-of-therapy-can-help-veterans-adjust-to-civilian-life/

Slide 26, military.com, Want To Support A Veteran Who Is Going Through A Tough Time?, Unknown date, retrieved March 3, 2022, https://www.military.com/benefits/veteran-benefits/be-there-for-veterans.html

Slide 31, Pew Research Center, The changing profile of the U.S, military: Smaller in size, more diverse, more women in leadership, September 10, 2019, retrieved March 3, 2022, https://www.pewresearch.org/fact-tank/2019/09/10/the-changing-profile-of-the-u-s-military/

Slide 32, Americans Flag Express, Poly-Max American Flag (Extreme Winds), Unknown date, retrieved March 3, 2022, https://flagsexpress.com/buy-american-flag/poly-max/

DR

DAILY REPUBLIC
Solano County's News Source

FAIRFIELD-SUISUN CITY, CALIFORNIA

Jeff and Lynn Jewell have dedicated decades to veterans' needs, and are still working. (Robinson Kuntz/Daily Republic)

Vacaville couple proves to be Jewell of veteran causes

By Todd R. Hansen

VACAVILLE — Lynn Jewell swore she would not marry into the military. Then she met Jeff, her husband of 39 years. "My dad was in the military, and my dad was TDY (temporary duty) all the time," said Lynn Jewell, 66, who was born and raised in Fairfield. "And I always said I would never marry anyone in the military... and then I met him." The Jewells met in September 1979. They were married in April 1980. Her father, Robert Markle, served in the U.S. Army Air Corp, then stayed in the Army after World War II. "He was shot down and had two Flying Cross medals before he was 21 years old," Jeff Jewell added to the discussion. But it was more than her father being away so much. Lynn Jewell now knows he suffered from some sort of post-traumatic stress disorder and was an alcoholic. It left her with a bad impression of the military lifestyle.

Jeff Jewell, 65, did not exactly love the Air Force when he chose to leave the service. He enlisted in 1973 and left in 1983. That included two tours at Travis Air Force Base. He and Lynn were scheduled to be married when he received orders to go to the Philippines. He tried the normal channels to change assignments, but when, to his disbelief, that failed, he broke chain of command and found someone else who could serve as the flying crew chief on that C-5. "My primary job was crew chief on a C-5, but I worked in all areas of maintenance . . . and I was a flying crew chief for almost two years," he said. Jewell called the duty "kicking the tires," meaning a full inspection of the gigantic cargo plane, which, by the way, has 28 tires. His missions included supplying the shah of Iran, Mohammad Reza Pahlavi, with aircraft and other military equipment before the Iranian Revolution led by the Ayatollah Ruhollah Khomeini. Iran had actually helped fund the development of the C-5 Galaxy. Jewell was not treated particularly well after he opted out of the Philippines assignment, canceled his re-enlistment and left the Air Force without regret.

Ironically, for two people who had no real desire to be around the military, the couple have since dedicated much of their lives to improving the lives of thousands of veterans. One of those veterans, as it turned out, was Jeff Jewell's father. Jim Jewell served during World War II as a sonar and depth charge specialist on a destroyer. "I don't know if he saw any action, but I know he couldn't hear," Jewell said. "And I filed a claim for my father and got him hearing aids." His grandfather on one side served on hospital ships during World War I. The other was in the trenches, and prior to that, he fought in the Mexican Border War.

Life of service stamped into Jewell.
Jewell was born in Rhinelander, Wisconsin, and at age 10 moved with his parents and five siblings to Green Bay and lived about three blocks from Lambeau Field. "I learned to drive in the parking lot of Lambeau," he said. It was during his high school years that a recruiter, who reminded Jewell of Jack Webb, signed him up under the Delayed Enlistment Program – a decision he kept secret from his parents even as they pressured him to make a decision about

college. Jewell said it was actually a trip he took as an Eagle Scout that opened his eyes to what existed beyond, what was at the time, the monocultural Wisconsin.

"I traveled all over Europe… and stayed with families and it just opened my eyes to there was life beyond Green Bay," Jewell said. The Air Force held the same promise of adventure, and largely delivered. After his service, Jewell took a job with the U.S. Postal Service, first in Pittsburg for a year, and then in Fairfield for more than four years. "And I didn't really think about helping veterans then," Jewell said. But when tensions grew between labor and management, Jewell helped develop an employee involvement program – the central pillar of which is something Jewell would take with him into what became his life's work with veterans. Jewell said it was clear in the postal dispute that no one was talking to each other, and even more necessary, no one was listening. He was named the Shop Steward of the Year for the East Bay in 1987. Then Jewell re-injured his back, and at the age of 33 found himself back in school – using veterans benefits he had known little about up to that point: the GI Bill and the Vocational Rehabilitation Program.

Jewell holds an associate degree in business from Solano Community College and a bachelor's degree in human relations from Golden Gate University, classes for which he attended at the former satellite school on Travis Air Force Base. He also has two master's degrees: one on public administration, and the second in psychology, which he studied for so he could serve as a Veterans Affairs counselor. "Those were some crazy years," he said. Lynn Jewell has a degree in accounting, and it was her work over the years that the couple relied on to survive. Solano County Veteran Services Office opportunity changes life It was while Jeff Jewell was at Solano College that a job as a benefits counselor at the Solano County Veteran Services Office came up. "I was the runner-up; I didn't get the job," Jewell said. But he continued his veterans work at the county vets office and at Solano College under a VA program.

"And that is when I started to hear the stories from the Vietnam veterans," said Jewell, who added that many were angry and feeling isolated. "And I never judged them. . . . I always tried to put myself in their own shoes and say, 'Wow, if I had all this going on, I'd be mad at the world, too,' " Jewell said. Not long after that, he got a call from Edna Barns, director of the county Veteran Services Office, who asked if he still wanted the job for which he had applied. When he told her yes, she hired him, and he worked for the county for 10 years, building a reputation as a fighter for veterans benefits and rights. "I was in Vallejo and I was supposed to handle all the claims for Vallejo and Benicia, but (veterans) would talk to me . . . (and in time), people would drive in from Vacaville and Fairfield," Jewell said. He did not like it when he was told to stop that, and soon after, left the county job and worked his way up to become the top administrator at the Veterans Affairs office in Concord, a position he retired from in October. But it was his work with a stand down for homeless veterans in the East Bay that connected him to the VA job. "I took a leap of faith, and it is the best move I made," Jewell said.

Lynn Jewell, who worked for Nut Tree for 13 years before taking a job with a local accounting firm, learned from her mother at an early age about the value of volunteering. "My mother was very active and volunteered a lot when I was growing up, so I volunteered with her," Jewell said. But it was not until the veterans hall in Vacaville needed someone to run the bingo program, and later manage the hall itself, that she became involved with veterans. Then as the list of volunteer activities and veteran group associations Jeff Jewell involved himself in grew, her commitment to the veterans grew, too. Among those activities is the annual Stand Downs for homeless veterans held at the grounds of the Dixon May Fair. Jeff Jewell helped start the stand downs in 2002. "I love what we do," Lynn Jewell said. She started in the background, but is now a point person for the event. The Jewells said among the work they do now that they are quite proud of is raising money for the Gold Star families who have lost a son or daughter in the military. They called it some of the most rewarding work they have ever done. That includes Jeff Jewell swimming from Alcatraz to Aquatic Park one year, raising $5,000 and more than $22,000 over the three-year period. "I was the last one in, but I made it," Jewell said. "I was standing on the beach absolutely terrified," Lynn Jewell added.

So many saved; important ones lost
Lynn Jewell said when she thinks back on the work they have done, she quite often thinks of all the volunteers who have helped along the way, how many became close friends, and now many are no longer with them. "We just lost another member of our organization. We just went to her funeral on (Dec. 12)," Jewell said. "And she was the wife of a Vietnam veteran." Despite the thousands of veterans the Jewells have helped, and in some cases, literally saved their lives, it is one that recently committed suicide that Jeff Jewell feels most closely. The man had helped Jewell in his professional journey, but had become homeless in the Santa Barbara area. He hanged himself in a public park. "That's why it hurts so much... I couldn't save him."

Printed in January 05, 2020 edition on page B1. Last Modified on January 3, 2020 at 3:34 pm.

News Release

November 6, 2019

Contact: Susan Shiu, PIO, 925-313-1183
Susan.Shiu@contracostatv.org

Board of Supervisors Honors Veterans and Invites You to the 101st Annual Veterans Day Celebration November 12, 2019

(Martinez, CA) – Contra Costa County's Board of Supervisors honors veterans and invites the community **to the 101st Annual Veterans Day Celebration on Tuesday, November 12, 2019, at 11:00 am** in Board Chambers at 651 Pine Street, Martinez, CA.

This free celebration event will feature a color guard ceremony, a special poetry reading, and keynote speaker Jeffrey Jewell, U.S. Air Force Veteran who recently retired as the Director of the Concord Vet Center. He has served veterans across Contra Costa County and throughout Northern California for nearly 20 years.

"On this Veterans Day, we express our profound gratitude for the service and sacrifice of our veterans," said Board Chair, Supervisor John Gioia. "We honor these heroes who protect the ideals of freedom and democracy."

"The Board of Supervisors thanks all veterans and their families, and recognize the work of veterans' services organizations, including the County's Veterans Service Office," Gioia said.

Following the Veterans Day Ceremony in Board Chambers, the public can continue the celebration at the Martinez Veterans Memorial Building at 930 Ward Street (corner of Court Street and Ward Street) in Martinez for lunch.

Shuttles for free transportation to and from the event at Board Chambers in Martinez will be available in east, west and south Contra Costa County. Please call (925) 313-1481 for shuttle times and details.

###

The Board of Supervisors of
Contra Costa County, California

In the matter of:

Resolution No. 2019/621

honoring Jeffrey Jewell the Concord Vet Center Director on his retirement

Jeffrey (Jeff) Jewell is a United States Air Force Veteran from 1973 to 1981 with two tours at Travis Air Force Base in California and one tour at Yokota Air Force Base in Japan as a crew chief on C5-A's; and

Whereas, prior to coming to the Department of Veterans Affairs, Jeff was a Veterans Benefits Counselor for ten years in Solano County; and

Whereas, he joined the Department of Veterans Affairs, Vet Center Team in 2001 at the Oakland Vet Center as Readjustment Counseling Therapist and transferred to the Concord Vet Center in 2003; and

Whereas, in March 2010, he was promoted to the director of the Sacramento Vet Center; and

Whereas, in January 2014, Jeff returned to the Concord Vet Center as the Director; and

Whereas, he was the lead bereavement counselor for the Vet Centers and he has been as the lead counselor at the Marine Memorial Annual Connection and Sharing Event with Gold Star Families for 16 years; and

Whereas, Jeff has been doing outreach at California State Prison Solano, Vacaville and California Medical Facility, Vacaville for the past 17 years to veterans that are incarcerated; and

Whereas, he has earned the following degrees all with honors: Associate of Arts in Business, Solano Community College; Bachelor of Arts in Human Relations, Golden Gate University; Master's in Public Administration, Human Resource Management, Golden Gate University; Master of Arts, Psychology, Marriage and Family Therapy, Chapman University; and

Whereas, Jeff is currently the 5th District Commander for the American Legion; and

Whereas, he is part of the Veterans Court in Solano County; and

Whereas, Jeff is the past chairman of the Vet Center National Homeless Veterans Working Group and he is the Director of the annual North Bay Homeless Veterans Stand-down.

that the Contra Costa County Board of Supervisors recognizes Jeffrey Jewell on the occasion of his retirement and honors his hard work and dedication to the veterans of our community.

Assemblyman Jim Frazier Honors Jeffrey Jewell as "Veteran of the Year"

written by ECT Jun 28, 2013

Anytime that a Veteran is honored, its a great day! This week, Assemblyman Jim Frazier (D-Oakley) named Jeffrey Jewell of Vacaville as his 2013 Veteran of the Year in the Assembly District. Here is the Press Release via the Assemblyman's Office.

Sacramento, CA – Assemblymember Jim Frazier (D-Oakley) has named Jeffrey Jewell of Vacaville as his "2013 Veteran of the Year" for the 11th Assembly District.

"Mr. Jewell's contributions to the welfare and improvement of the veterans' community are invaluable," said Assemblymember Frazier. "His exemplary record serves as an excellent model for all public-spirited people of the state, and we are grateful for all of his selfless contributions and sacrifices."

Jewell's commitment to community service was sparked at a young age when he joined the Boy Scouts of America, serving all the way through Eagle Scouts and Order of the Arrow. After graduating from high school, he joined the United States Air Force and simultaneously worked as an avid volunteer for the Suicide Prevention Hotline, Big Brothers, Boy Scouts, and also as an English teacher to Japanese businessmen while he was stationed in Japan.

Jeff began giving back to the veterans' community while working at the Solano County Veterans Service Office, which later lead him to become active with the Vacaville Veterans Building. His civic-service record speaks for itself; he served as past commander for Vacaville's Disabled American Veterans, the American Legion, and the United Veterans Memorial Association. He currently serves in various veterans' organizations including as Director of the Sacramento Vet Center, where he helps provide counseling, outreach, and referral services to veterans in the Sacramento area in order to help them make a satisfying post-war readjustment to civilian life. He also serves as Chair of the Vet Center National Homeless Veterans Working Group, and Co-Director of the annual North Bay Homeless Veterans Stand Down at the Dixon Mayfair Grounds.

"I am humbled and honored that Assemblymember Frazier would think to choose me," said Jewell. "I don't do this for recognition or a pat on the back; I do this because I truly love the work."

The California State Assembly Veterans Affairs Committee held an annual luncheon in Sacramento on Wednesday, June 26, 2013 to honor our service men and women. Attached is a photo of Assemblymember Frazier recognizing Jeffrey Jewell on the Assembly floor with an Assembly Resolution.

To contact Assemblymember Jim Frazier please visit his website at http://www.asmdc.org/members/a11/ or call his District Office at 707-399-3011.

Follow Assemblymember Jim Frazier on Facebook and "Like" him for updates on events and happenings in the 11th AD.

ECT
Publisher of EastCounty Today and host of several Podcast Shows

THE REPORTER

Kelli's Heroes: Jeff Jewell to speak at Memorial Day ceremony

By THE REPORTER |

Memorial Day (observed) is Monday May 28 and will be observed in a ceremony hosted by the Vacaville Veterans. The ceremony will take place starting at 11 a.m. at the Vacaville-Elmira Cemetery. This year the key note speaker is a veteran who works hard every day to 'care for those that bore the battle, their wives and orphans."

In the truest sense of exactly how veterans and their families should be helped, you will find Jeffrey Jewell. Jeff is a veteran of the US Air Force where he served as a crew chief on C5-As. Since that time he started out as a benefits counselor to veterans with Solano County, and from there has moved into positions within the US Department of Veterans Affairs.

He currently is the Director of the Concord Vet Center. He is the lead bereavement counselor for the Vet Centers and has been doing bereavement counseling with Gold Star Families for more than 15 years. His outreach and care for veterans and their families and his perseverance in ensuring the right things are done to help our veterans is the only way I can describe the passion this man displays to serving his fellow veterans. You won't want to miss what he has to say, so bring your lawn chairs and join us on a day when grateful nation pays tribute.

The author is a local advocate for veterans' issues. Email: vetsfrst@aol.com.

PUBLISHED: May 18, 2018 at 12:00 a.m. | UPDATED: August 29, 2018 at 12:00 a.m.

Legion Post Supports Travis Air Force Base Fisher House

by Jeff Jewell | Aug 11, 2021

We had another successful delivery of an evening meal for 20 residents of the Travis, Air Force Base Fisher House on Monday, the second Monday of the month. This is something our Legionnaires starting doing three months ago at Post 165 in Vacaville. Thus far, we have delivered meals to 57 residents on the first Monday of every month at the Fisher House on Travis, AFB. Thank you to American Legion Member Michelle Hammer-Coffer, pictured below having recently broken her arm, we decided help her out. Michelle paid for the food and the meal was prepared by American Legion Member "EE" who is a retired Army chef and former chef at VA Medical Center in San Francisco. Jeff Jewell, post adjutant, washed all the dishes at the Veterans Memorial Hall kitchen where the meal was prepared. Jeff, also delivered the meal to the Fisher House. The meal was grilled/baked pork chops in a special sauce with butter, garlic and brown sugar, fresh greens beans with onions and bacon, au gratin potatoes and homemade pineapple up-side-down cake. If you are interested in providing a meal on the second Monday of the month in 2022 let me know by email.

Mike Goble, center, receives a check from Vacaville's American Legion Post 165 members, from left, Chaplain Dan Seibert, Cmdr. Michael Terhorst, Vice Cmdr. Leonard Miller and Adjutant Jeff Jewell, Friday, March 25, 2022. (Robinson Kuntz/Daily Republic)

Vaca veterans contribute to Memorial Hill flag fund

By Todd R. Hansen

VACAVILLE — Mike Goble never imagined putting up a couple of flags on a hill in memory and honor of two friends who died in Vietnam would ever become what it has turned out to be. "It's become a kind of landmark, way bigger than what I ever thought it would be," Goble said about the flags that fly on what is now known as Memorial Hill. "And it is one of the best views in Solano County," he said.

The Vacaville veterans, and American Legion Post 165, presented Goble with a $380 check Friday to cover the cost of a "round of flags" that now fly on the hill. "I have had previous encounters with members of that post, and I was called by one of the members and he asked me how much for the flags," Goble said. There are the original two U.S. flags, and the corresponding service flags for Marine Cpl. Richard Strahl, who died in a mortar attack in 1967, and Army Capt. John Curran, who was killed in 1971. Goble, who served in the Marine Corps, went to high school with Strahl and Curran in Glendale, Arizona.

There is also an American flag for CHP Officer Kirk "Hollywood" Griess, who was killed while on duty in August 2018 and served in the U.S. Marine Corps before joining the California Highway Patrol. "I had met him earlier (that) year and found out he was a Marine," said Goble, who also had played Little League baseball with Brett Machado, who married Griess' daughter, Kadi. She is a Fairfield police officer. Goble later took Griess' parents up the hill, folded and presented to them the original flags placed there in honor of their son. There also are a Purple Heart flag, a POW/MIA flag, service flags for the military branches and a series of fallen officer and fallen firefighter banners on the hill, along with two dedicated benches and some plaques. "It just keeps getting bigger," said Goble, who is happy to add another in honor of others who died in service of their country in the military, or had served their communities in the police or fire departments. "So I tell people to give me a call and I will (put up) a flag for them, too," Goble said.

The connections do not necessarily have to be local. For example, he honored the memory of Natalie Corona, who was a Davis police officer from Arbuckle, who was killed on duty in January 2019 while responding to a traffic accident. Goble said he and a couple of members of the Vacaville veterans went up to Memorial Hill and changed out some flags as well. Of course, the flags honoring his friends touch him the most, the first going up just before Memorial Day in 2017, but the feeling of adding another flag does not change for Goble. "I'm glad to do it," he said.

• To add a flag, or if interested in donating, contact Mike Goble on his Facebook page.

Printed in the March 27, 2022 edition on page A3. Last Modified on March 26, 2022 at 9:29 am.

Jeff Jewell gestures to memorabilia and other items held at the Vacaville Veterans Hall, on Wednesday, June 19, 2019. The hall was recently renovated.(Robinson Kuntz/Daily Republic)

Vacaville veterans hall completes $1.5M renovation

By Ian Thompson

VACAVILLE — The 90-year-old Vacaville Veterans Memorial Hall has restored its main room's original vaulted ceiling as the centerpiece of a $1.5 million renovation project that is just wrapping up. "We needed to restore this hall back to its original luster," said Jeff Jewell of American Legion Post 165.

The veterans will celebrate the end of a nine-month-long renovation with a rededication ceremony at 2 p.m. Sunday at the building, which is located at 549 Merchant St. The open house will continue to 6 p.m. They have invited Rep. John Garamendi, D-Walnut Grove, state Sen. Bill Dodd, D-Napa, state Assemblyman Jim Frazier, D-Discovery Bay, members of the Solano County Board of Supervisors and county officials, Vacaville officials, local longtime volunteers and community leaders. "We want to show off the building," said Lynn Jewell, secretary of the Veterans Memorial Building Association.

Workers are now putting finishing touches to the new front doors and entryway. They have already replaced the roof and HVAC systems, restored the hardwood flooring, improved the front and rear entryways, upgraded refrigeration equipment and repaired collapsed French drains. The false ceiling in the upstairs hall that covered the original vaulted ceiling since a previous 1992 renovation has been removed and the wood beams have been repaired. The $650,000 1992 renovation was paid out of veterans groups' pockets, but this renovation was funded by Solano County in its campaign to fix up the county's veterans halls in Dixon, Benicia, Suisun City and other locations. A new trophy case for the Brotherhood of Vietnam Veterans has been added, and other veterans groups will be able to put up documents, photos and memorials in the renovated annex area.

The building houses four veterans' organizations and two auxiliaries, and is a meeting place for Boy Scouts troops, charity fundraisers, community events, patriotic celebrations and holiday dinners for the community. Veterans moved their annual Thanksgiving and Christmas community dinners to the Three Oaks Community Center last year while the renovation was underway. It also provides a venue for wedding receptions, birthday parties, baby showers, quinceañeras and various other celebrations. The hall's July poker tournament will be the first event to enjoy the buildings improvements, Jewell said.

Printed in the June 20, 2019 edition on page A1. Last Modified on June 19, 2019 at 9:10 pm.

(Robinson Kuntz/Daily Republic)

A Christmas Eve meal delivered

By **COREY KIRK** | ckirk@thereporter.com |

Members of the Vacaville Veterans and volunteers from the community prepped and delivered meals to those in need in Solano County. In the early morning hours of Christmas Eve, Adjutant of the American Legion and Auxiliary Post 165 Jeff Jewell turned the lights on at the Vacaville Veterans Memorial building on Merchant Street. With nearly 1,000 meals to be prepped, a long day of serving the community was ready to commence. "It touches my heart," Jewell said about serving others.

The Vacaville Veterans comprises five other organizations with American Legion Post 165: American Legion Auxiliary Unit 165, America Veteran's Post 1776, Disabled American Veterans Chapter 84, Veterans of Foreign Wars Post 7244 and Veterans of Foreign War Auxiliary Unit 7244. For the past nearly three decades, the Vacaville Veterans have hosted meals on Thanksgiving and Christmas Eve. Preparations for the event begin nearly a week before, when they properly clean and sanitize the kitchen. "It takes the whole day, about 20 people," Jewell said.

After the area is ready to go preparing the food begin, from the house-made bread to the carving of the endless turkeys and hams with Army veteran chef Exequiel Enriquez leading the way. As the sun rose, volunteers arrived with a desire to help and get the meals ready for delivery. Among them was Senior Master Sgt. Reynaldo Rios, who serves at Travis Air Force Base. He was excited to return to help after doing so last year. A member of the Air Force Sergeant Association who works closely with the American Legion, he was happy to do his part. "As an active duty member, we tend to really look forward to not just serving our nation but our local community," Rios said. "What that does is that brings the satisfaction to the overall team that we are all in this together. We have all had a pretty rough 15-16 months and knowing that we can all still come together and take care of those that are less fortunate, that really makes it known we are a part of something bigger than ourselves."

Once all the volunteers arrive, plating proceeds in an assembly line. Hours later, drivers arrived at the building ready to deliver the meals. The mission of the overall day was to distribute meals — mainly to community members living in underserved areas or veterans in need of a meal. Among the drivers were brothers Adam and Andrew Grabowski, who have been working this

event for nearly a decade but began doing the delivery process once the COVID-19 pandemic began. "We try to find the best ways to help but also making sure that we are safe and our community is safe," Adam said. Although Adam is finishing his fourth year at UCLA and Andrew lives in Washington, DC with a career in aerospace engineering, they love coming back to their hometown to help. "Even though we are all in different places, giving back to this it's a good family tradition, it's a good way to give back to the community," Andrew said.
Delivery drivers were given a set of addresses across Solano County. Among the many stops for the Grabowski brothers was the home of Jim Noblitt, who was excited to try his first meal from this event. "My kids are all at other commitments so this really helps," Noblitt said.

Later on their delivery run was Bill Lebar, an Air Force veteran who has received the Christmas Eve meal dinners for four years. Now that his wife, Helen is unable to drive, he was excited to dig into the hot meal. "You just can't beat it," Lebar said. "We just feel really blessed that we have this program." Air Force Veteran Bill Lebar is all smiles when he had his meal delivered early Christmas Eve afternoon.

Over the span of just a few hours, volunteers were able to deliver hundreds of meals. "When you come to an event like this, you see there is so much more love and care in our world then you see on TV," Rios said. "When you come to this event it's full of selfless individuals that are really looking out for everyone other than themselves." For those interested in supporting the holiday meal programs, donations can be made to https://www.vacavets.org/free-holiday-dinners/.

For more information on other events, visit vacavets.org.

PUBLISHED: December 24, 2021 at 5:27 p.m. | UPDATED: December 24, 2021 at 6:28 p.m.

Yaretzi Peina visits with Santa Claus before receiving gifts at the Vacaville Veterans Community Christmas Dinner at the Three Oaks Community Center in Vacaville, Monday, Dec. 24, 2018. (Robinson Kuntz/Daily Republic)

Veterans serve up holiday Christmas Eve cheer in Vacaville

By Susan Hiland

VACAVILLE — The smell of turkey wafted Monday throughout the Three Oaks Community Center during the annual Vacaville Veterans Community Christmas Dinner. Jeffrey Jewell of American Legion Post 165 has helped out with the dinner since 1986. For the past few years he has been the man wearing the name tag, "Boss." "All the veterans groups work together to put this on," he said.

The preparations began at 5 a.m. Christmas Eve with 200 volunteers boxing up food for various communities, including Fairfield and Travis Air Force Base. "We deliver as Meals on Wheels today," Jewell said. The volunteers take 40 to 50 meals to the homes of seniors in the area along with meals to those who called in with requests for a delivery because they're homebound. The numbers of the donated food is impressive with 50 turkeys, 270 pounds of ham, 60 trays of yams and pounds upon pounds of corn and mashed potatoes. "The dressing is homemade at the Veterans Hall during the week," Jewell said.

On the stage were tables stacked with toys for children from ages 0 to 18 years old. Santa Claus came at noon and handed out the packages. Jerry Kirpatrick of Vacaville was not a youngster but he did come for a nice, home-cooked meal. "It's nice to have a good meal for a change since I'm a bachelor," he said. Jack Gardner, 19, came home for Christmas from UCLA and for 12 year in a row has helped his father, Rory Gardner, fix and serve the meals. The rest of the family was coming later to help out, he said. "I want to support the veterans since my grandfather was in World War II," Jack Gardner said. "I thought it is important to give back because they gave so much." He was joined by Linda Bashqoy from Petaluma, who is in the Coast Guard. "A friend told me about this event and I wanted to help," she said. Since Bashqoy's family is 3,000 miles away, it was something to do for the holidays. For the past three years William Reed of Vacaville has helped cook and serve the meals. As a child, his family struggled and came for meals at the center on Christmas Eve. He also has helped with passing out presents to the children. "It's a way to give back and my mother Linda is at the front door helping people get signed in," Reed said. "We haven't forgotten how this helped us."

Volunteers serve food at the Vacaville Veterans Community Christmas Dinner at the Three Oaks Community Center in Vacaville, Monday, Dec. 24, 2018. (Robinson Kuntz/Daily Republic)

Printed December 26, 2018 edition page A3, Last Modified on December 25, 2018 at 12:01 am.

https://youtu.be/9xHPQMYROcA

![Veterans' Voices logo]

Contra Costa County
CCTV• 10 Douglas Drive Suite 200 • Martinez, CA 94553 •
www.contracostatv.org/veteransvoices

May 5, 2022
MEDIA RELEASE

Contact: Nathan Johnson, (925) 313-1481
Nathan@vs.cccounty.us

Veterans Voices Show to Discuss Veterans and Social Isolation

On **Monday, May 9th at 7 pm**, Veterans Voices will host a live panel to discuss the many ways in which Veterans are prone to social isolation and provide real life solutions to help them reconnect. Guests will include Jeff Jewell, Harry Hitchings, and Jet Garner who have spent their careers helping Veterans and/or advocating for outdoor therapy for Veterans. They will be joined by our host, Nathan Johnson, to have an in-depth conversation of the causes and solutions for social isolation in the Veterans' community.

Viewers can share their own experiences or ask a question by calling into the program at (925) 313-1170. They can also use the phone number to record a message for the panel any time prior to the show. Leave questions and comments at facebook.com/veteransvoices1 or email veteransvoices@contracostatv.org prior to or during the live television show.

The Veterans' Voices program is broadcasted live on Contra Costa Television (CCTV) on the second Monday of every month at 7 pm. It can be seen live online at facebook.com/veteransvoices1. You can find an archive of past shows and show resources at contracostatv.org/veteransvoices or on our YouTube channel Veterans Voices of Contra Costa. The show is re-run multiple times on CCTV on Comcast Channel 27, Astound Channel 32, and AT&T U-Verse Channel 99.

So many who serve find themselves isolated after transitioning from military to civilian life. It is our hope that this show will reassure Veterans that they are not alone and point them in a positive direction with concepts and resources to help them reconnect.

###

Help The Cause

E5 Therapy was incorporated as a nonprofit organization in the State of California on April 2, 2020. In July we applied to the IRS for 501(c)(3) certification and were approved Oct 2nd, 2020.

You are now welcome to use PayPal to donate to E5 Therapy. If you prefer to send a check, you can mail it to:

E5 Therapy
333 Sunset Avenue, Suite 110
Suisun, CA 94585

100% of all donations received are used to **directly** subsidize the cost of our counseling and training programs for our Veteran clients.